a first
reader in
statistics

second
edition

a first
reader in
statistics

second
edition

freeman f. elzey
san francisco state university

brooks/cole
publishing company
pacific grove, california

a division of
wadsworth, inc.

Brooks/Cole Publishing Company
A division of Wadsworth, Inc.

ISBN: 0-8185-0140-5
L.C. Catalog Card No.: 74-83225
Printed in the United States of America
20 19 18 17 16 15 14 13 12

Production Editor: Lyle York
Interior & Cover Design: Linda Marcetti
Illustrations: Creative Repro Photolithographers,
    Monterey, California
Typesetting: Reprographex, Palo Alto, California
Printing & Binding: Malloy Lithographing, Inc.,
    Ann Arbor, Michigan

preface

      Like the preceding edition, this book is an introduction to the rationale and fundamental concepts upon which many statistical analyses are based. It is written for those who wish to gain an understanding of statistical concepts without becoming involved with the actual application of the techniques. No symbols, formulas, or mathematical derivations are used. This edition provides an expanded discussion of many of the concepts covered in the previous edition; particular attention is given to the use of confidence intervals and comparisons among multiple samples.

      Although many beginning statistics texts explain the fundamentals of statistical analysis very clearly, most books attempt to present all at the same time the theoretical basis of statistics, mathematical derivations of formulas, computational formulas, and applications of the various techniques. Faced with this many-faceted approach, the beginning student often becomes so absorbed in the mechanical manipulation of symbols and equations that he loses sight of the underlying basis of statistics. He may come out knowing how to substitute numbers for symbols and solve equations but having little or no understanding of the theory that underlies them.

In this book, we shall work our way through the basic rationale upon which almost all common statistical analyses are based. The reader will not be concerned with statistical formulas or with the numerous types of statistical techniques. Instead, I shall emphasize the ways in which statistics are used to describe data and make inferences from samples.

For their helpful comments and suggestions for the first edition of this book, I would like to express my appreciation to Professors Frederick B. Davis of the University of Pennsylvania, Raymond G. Johnson of Macalester College, Charles R. Kessler of Los Angeles Pierce College, Edward Nolan of the University of New Mexico, O. J. Rupiper of the University of Oklahoma, E. I. Sawin of San Francisco State University, and Ezra Wyeth of San Fernando Valley State University.

I would also like to thank Professors Malcolm D. Arnoult of Texas Christian University, Homer H. Johnson of Loyola University of Chicago, Garvin McCain of the University of Texas at Arlington, Janet Moursund of the University of Oregon, and Robert S. Steele of Wesleyan University for their helpful reviews of the manuscript for the present edition.

Freeman F. Elzey

contents

a first reader in statistics

second edition

1

organization and description of data

We'll begin with a consideration of what research data
are and how we can organize them. Basically, a set of statistical
data consists of numbers that represent measures of some
property or phenomenon. To be useful in statistical analysis,
the data must consist of different numbers; that is, they must
not all be of the same value. The numbers may represent
achievement scores, physical measurements, the number of
people selecting certain categories, and so forth. We shall be
concerned with statistical methods appropriate to a variety of
types of data.

Let's consider one of the most common types of data
collected by researchers—test scores obtained from a number of
individuals. Suppose we give an arithmetic test to a group of
11 fourth-grade students and obtain the following test scores:

9, 12, 15, 13, 15, 15, 17, 8, 11, 14, 12.

By looking at these scores we can get some idea of how our
group as a whole scored on the arithmetic test. If we had tested

a very large group of children, however, it would not be easy to get a total picture of the group of scores without first organizing them in some fashion.

frequency distributions

Sometimes the most useful way to organize a set of scores is to make a *frequency distribution* of them. A frequency distribution is made by placing the score values in numerical order, from the largest down to the smallest, and indicating the frequency with which each value occurs. It is customary to place the largest score at the top of the distribution. For our arithmetic scores, we see that the largest score value is 17 and the smallest is 8, so our frequency distribution will encompass score values from 17 to 8, inclusive. Below is the frequency distribution of our arithmetic scores.

Score Value	Frequency
17	1
16	0
15	3
14	1
13	1
12	2
11	1
10	0
9	1
8	1
Total number of scores	11

Notice that the "Score Value" column contains all possible score values from 17 to 8, inclusive. The "Frequency" column indicates how many people received each of the arithmetic scores. In other words, it represents the frequency of occurrence of each score value. Checking this column against our original set of scores, we find that there are indeed three children with a score of 15, none with scores of 16 or 10, and so on. Also, the sum of the "Frequency" column should equal the total number of children who were tested; and it does total 11.

One important thing to notice in this frequency distribution is that each score value is listed, regardless of whether anyone received it. In our data, 16 and 10 are listed as having frequencies of zero, since no children obtained these particular scores.

measures of central tendency

Now that we have the data organized in a frequency distribution, we need some way of conveniently describing the entire distribution. One way is to determine the most "typical" score and use it to describe the total group of scores. In statistics there are many kinds of "typical" scores, and they are referred to as *measures of central tendency;* that is, they represent points in the frequency distribution around which the scores tend to center.

One of these measures of central tendency is the *modal score.* It is usually referred to simply as the *mode* of the distribution. The mode is easy to locate in a frequency distribution because it is merely the score value that has the largest frequency. In our data, the mode is 15 because more children received that score than any other score. The mode is not very helpful in describing the entire frequency distribution, however, because it only tells which score was received most frequently. If we know only the value of the mode, we know nothing at all about the values or relative positions of any of the scores.

If two or more score values have the same frequency, and this frequency is larger than that of any other score value, the distribution has more than one mode, and we say it is *multimodal.* For example, if two score values have identical frequencies, and the frequencies are larger than the frequency of any other score value in the distribution, the distribution is *bimodal.*

Another measure of central tendency is the *median.* The median is the score value that marks the midpoint of the set of

data. Fifty percent of the scores are above the median and fifty
percent are below it. In our distribution, the median is 13
because five people have scores above 13 and five have scores
below 13.

The median is generally a bit more descriptive of the
frequency distribution than the mode because it gives us the
middle score of the distribution, whereas the mode tells us
nothing about how the scores are distributed. However, the
median does not take into account the *values* of the scores
above or below it. It is determined only by how many scores
there are on either side.

A more exact measure of central tendency, which takes
into account the value of each score, is called the *arithmetic
mean*. The *arithmetic mean* of the distribution is what is commonly called the *arithmetic average* of all scores. I shall refer
to it simply as the *mean*. It is obtained as you ordinarily obtain
averages, by adding all the scores together and dividing by
the number of scores. The sum of all the arithmetic scores in
our set is 141. Dividing this sum by 11, the number of children,
we obtain 12.8, which is the mean of the distribution.

The value of the mean is computed by using the *value* of
each score in the distribution, whereas neither the mode nor
the median is based on the values of scores; the mode is based
solely on the frequency of scores and the median is based on
the relative positions of the scores without regard to their values. Ordinarily, the mean is the most useful and stable measure
of central tendency. For most sets of data, the mean is the
most representative score. Furthermore, it lends itself to a wide
variety of statistical manipulations; the other two measures
do not.

In summary, we have learned how data are arranged in a
frequency distribution, and we have defined three measures of
central tendency:

> *Mode.* The most frequently occurring score in a frequency dis-
> tribution.
> *Median.* The point in a frequency distribution below which
> half of the scores lie.
> *Mean.* The arithmetic average of the scores in a frequency dis-
> tribution.

measures of variability

To describe a set of data adequately, we need both a measure of central tendency and a measure of variability. If a researcher tells us that the mean score in a set of data is 12.8, we have some information about the scores of the group, but we have no indication of how the scores are spread among the individuals; that is, we have no information regarding the *variability* of the scores.

The crudest measure of variability is the *range*. The range is simply the number of score values encompassed by a frequency distribution. The scores in our set range from 8 to 17. Therefore the range of scores in this distribution is 9, the difference between the highest and lowest values in the distribution. Obviously, the range merely tells us the span of score values in a distribution; it tells us nothing about the nature of the distribution of scores within the limits of the range.

Another, somewhat more useful, measure of variability is called the *interquartile range*. This is the range of scores that encompasses the middle 50% of the scores, leaving 25% above the interquartile range and 25% below it. For example, if we are told that a frequency distribution of scores has an interquartile range of 40 to 55, we know that the middle 50% of the distribution lies between these two score values and that 25% of the scores are below 40 and 25% are above 55. Unlike the range, the interquartile range is not affected by the value of the extreme scores in the distribution, and therefore it depicts the variability of the scores more accurately.

The *average deviation,* or *mean absolute deviation,* although rarely used, is a more refined measure of variability than either the range or the interquartile range in that it is computed by using the value of each of the scores rather than just their relative positions in the distribution. The average deviation is just what the name implies: the average amount that the scores deviate from the mean, regardless of the sign of the deviation. The first step in computing it is to determine the amount by which each score deviates from the mean. The average deviation is computed by summing the deviations of all the scores, without regard to whether they are positive or negative, and dividing this sum by the number of scores in the distribution. However, the average deviation is rarely used, because it does not lend itself to further statistical analysis and therefore has limited usefulness.

Probably the two most useful measures of variability are the *variance* and the *standard deviation*. The variance is computed by finding the deviation of each score from the mean, squaring each of these deviations, summing all of the deviations, and dividing the sum by the number of scores. The standard deviation is a measure of variability based on the square root of the squared deviations. That is, the standard deviation is the square root of the variance.

We shall first be concerned with the use of the standard deviation as an indicator of the degree of dispersion of a set of data. The use of the variance in statistical analyses will be taken up later. To better understand the usefulness of the standard deviation, we first need to look at some ways in which data are often distributed.

distribution curves

It is possible, of course, for data to be distributed into any shape. For instance, if you obtain a set of scores in which almost everyone received relatively high scores, the distribution will look something like that shown in Figure 1.

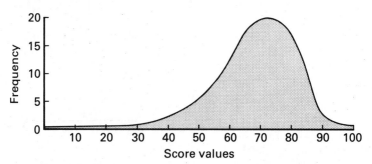

Figure 1. Distribution curve of a negatively skewed distribution.

The curve shown in Figure 1 is called a *distribution curve*. It is a graphic representation of a frequency distribution of scores, with the score values indicated on the horizontal axis and the frequency of each particular score value indicated on the vertical axis. In this particular distribution, 10 people received a score of 60, so a point is marked directly above score value 60 and across from frequency 10. The frequency for each score value is plotted in this manner and then a curve is smoothly drawn through the points to indicate the *shape* of the distribution.

The curve in Figure 1 is said to be *negatively skewed* because the *tail* of the distribution trails off to the left, in the direction of the lower score values. Twenty people were at the mode of this distribution. It is evident that a large majority of people received high scores on this test because the mode of the distribution is a value of 75, and most scores lie very close to that value.

Had it been a very difficult test, with most students receiving very low scores, the distribution would be *positively skewed* and would look like the curve shown in Figure 2, with the tail going off to the right, in the positive direction.

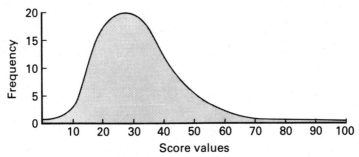

Figure 2. Distribution curve of a positively skewed distribution.

Sometimes data are distributed in such a way that there are two modes in the distribution. Such distributions, as mentioned before, are called *bimodal* distributions. A bimodal distribution curve is presented in Figure 3. Notice that the score values 30 and 70 both have a frequency of 15, which is the

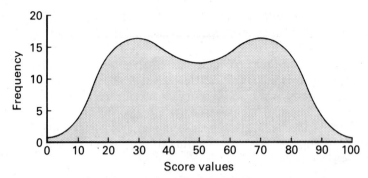

Figure 3. Distribution curve of a bimodal distribution.

largest frequency of any value in the distribution. Of course, it is possible for a distribution to have three or more modes, in which case it is described as a *multimodal* distribution.

The degree of variability of the data is shown in its frequency distribution by the breadth of the curve relative to its height. We can see in Figure 4 that there is less variability in

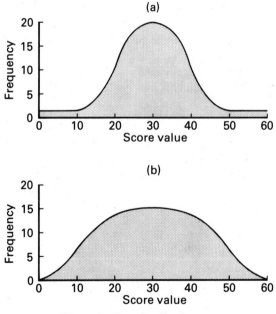

Figure 4. Two distribution curves

the data for curve (a) than in the data for curve (b). Had we actually computed the standard deviations of the data in these two distributions, we would have found the standard deviation to be 8 for distribution (a) and 12 for distribution (b). Thus we see that the larger the standard deviation of a distribution the more variability it has. The data in curve (a) are said to be less variable, or more *homogeneous,* than the data in curve (b).

The means of the two distributions in Figure 4 are the same; both are 30. This figure illustrates the fact that distributions having identical means may be very different in variability. Two measures are needed to describe a set of data: a measure of central tendency (the mean), which gives us a typical or representative value for the distribution, and a measure of variability (the standard deviation), which gives us an indication of the degree of dispersion among the data.

2

the normal distribution

As mentioned previously, distributions come in all shapes and sizes. Fortunately for the statistician, many distributions of naturally occurring data tend to follow a certain pattern called a *normal distribution*, especially when there are many scores involved. The curve that can be drawn for these distributions is called a *normal curve*. We shall spend some time examining the properties of the normal distribution, since much of the remainder of the book will be based upon these properties. Figure 5 presents the basic shape of the curve of a normal distribution.

As with all curves that are graphic representations of frequency distributions, this curve depicts the score values along the base line, or horizontal axis, and depicts the frequency of the scores along the vertical axis. The normal curve is theoretical and infinite and therefore is only approximated in actual distribution. Figure 5 shows the basic shape of all frequency distributions in which the measures are normally distributed. (Notice that the tails of this theoretical distribution never actually touch

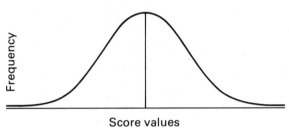

Score values

Figure 5. The normal distribution

the base line.) This curve could represent the IQ scores of all
eighth graders, the heights of all 44-year-old men, or the racing
times of all runners in the 100-yard dash. Since we are interested only in the properties of this curve, the actual frequencies
and score values are of no interest to us at this time.

A number of characteristics of the normal distribution
are of considerable value to the statistician. One characteristic is
symmetry: the distribution has the same shape on either side
of the center point, the largest frequency of scores being located
around the center and the smaller frequencies of scores occurring at the two tails of the distribution.

Another important characteristic of this distribution is
that its mean, median, and mode are identical. This fact indicates that the mean, or arithmetic average, is also the most frequently recurring score (characteristic of the mode) and lies
at a point that divides the distribution exactly in half, with 50%
of the scores lying above the mean and 50% lying below it
(characteristic of the median).

the standard deviation of a normal distribution

Chapter 1 introduced the concept of variability, using
the standard deviation as a measure of the degree of dispersion
among a set of data. The standard deviation takes on added
importance for statistical work when it is used as a measure of
variability of normally distributed data.

To illustrate the use of the standard deviation, suppose
we have a set of data that are normally distributed, as shown in

Figure 6. The mean, as in all normal curves, is located at the exact center of the distribution. To show graphically where the standard deviation is located on a normal curve, we can find the point on the curve at which it starts growing faster horizontally than it grows vertically—the *point of inflection* of the curve—and we can draw a perpendicular line from this point to the base line. This line is shown as a dotted line in Figure 6. If we calculate the value of this standard deviation in this normal distribution, we find that it indeed lies at the point of inflection.

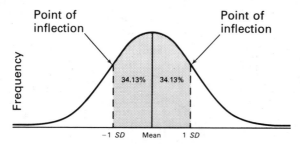

Figure 6. The normal distribution

Since the curve is symmetrical, what holds true on one side of the mean also holds true on the other side. Therefore we have two points of inflection and two vertical lines dropping to the base line from these points.

The total area under the curve represents all of the scores in the distribution, and it can be divided into parts representing percentages of the whole. I have already said that the area to the right of the mean contains 50% of the scores. Now it is a characteristic of the normal curve that other percentages can be determined using the mean as a point of departure. For example, 34.13% of the area under the curve lies between the mean and the vertical dotted line we have drawn. This area therefore contains 34.13% of the total area under the curve. The distance along the base line from the mean to the intersection with the dotted line is, by definition, the *standard deviation* of the distribution. The range of scores between the mean and one standard deviation encompasses 34.13% of the scores in the distribution. This, of course, holds true on both sides of the mean. Thus we have a point on the base line to the right of the mean that is termed "one standard deviation above the mean" or, simply, 1 *SD*, and also a point to the left of the mean that is termed "one standard deviation below the mean" or −1 *SD*. Both of these points are shown in Figure 6. We can see that 68.26%, or about $\frac{2}{3}$, of the scores in the distribution lie between 1 *SD* and −1 *SD*.

If we measure the distance along the base line from the mean to 1 *SD* and then measure the same distance along the base line from 1 *SD* toward the tail of the curve, we locate the point that indicates the second standard deviation, or 2 *SD*. The distance along the base line from the mean to 1 *SD* and the distance from 1 *SD* to 2 *SD* are exactly the same, even though the *areas* under the curve are different for these two portions. Again, this holds true on both sides of the mean, and we can locate 2 *SD* and −2 *SD* as shown in Figure 7.

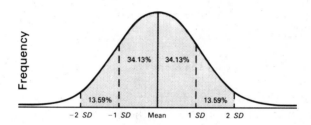

Figure 7. The normal distribution

We recall that 34.13% of the scores lie between the mean and 1 *SD*. However, it is clear from Figure 7 that there is a much smaller percentage of scores between 1 *SD* and 2 *SD* —only 13.59%.

We have positioned the point that designates 2 *SD* so that it is twice as far from the mean as 1 *SD* is. Likewise, we can locate a point along the base line that is three times as far from the mean as 1 *SD* and call this point 3 *SD*. This distance, too, can be located to the right and to the left of the mean, giving us 3 *SD* and −3 *SD*, as shown in Figure 8.

Figure 8 also shows us that a very small percentage of the scores lie between 2 *SD* and 3 *SD*—only 2.14%, in fact.

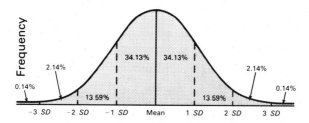

Figure 8. The normal distribution

The normal curve, being a theoretical curve, never touches the base line. For all practical purposes, however, the curve virtually touches the base line at 3 *SD*. Indeed, as Figure 8 indicates, only 0.14% of the scores lie above 3 *SD* and 0.14% below −3 *SD*. We can also see from Figure 8 that the total area under the curve accounts for 100% of the scores (the sum of all of the percentages shown).

Now let's examine how we can interpret any particular score if we know only how far, in standard deviation units, it is removed from the mean. To do this, we shall use a concrete example of IQ scores and assume that they are normally distributed. In the distribution presented in Figure 9, the mean IQ score is 100 and the *SD* of the scores is 15 points; that is, the value of the IQ that lies at 1 *SD* is 15 points larger than the value of the mean. We now have a measure of central tendency and a measure of variability. Since we assumed the shape of the distribution to be normal, these two measures are all we need in order to obtain a complete picture of the distribution.

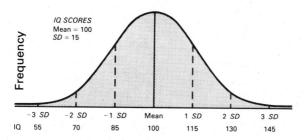

Figure 9. Normal distribution of IQ scores

We know that 50% of the IQ scores are above the mean IQ of 100 and that 50% are below this mean. With an *SD* of 15, we know that at 1 *SD* from the mean the IQ score is 115 and that at −1 *SD* from the mean the IQ score is 85.

Similarly, we can compute the IQ scores at the other *SD* points shown in Figure 9.

Using the percentages associated with the various *SD* units, we can determine any person's position relative to the total group by computing how far, in standard deviation units, his score is removed from the mean of the group. For example, suppose a person has an IQ of 115. We can easily tell that 84.13% of the total group have IQ scores below his and that 15.87% have IQ scores above his. Statistical tables are available that give the percentage of scores associated with any *SD* value or fraction of *SD* value in the normal distribution; therefore one can be very precise in describing the position of any particular score in *SD* units, such as 1.42 *SD*, −2.46 *SD*, 0.08 *SD*, and so forth.

In essence, one of the important properties of the normal distribution is that, knowing the mean and *SD* of the distribution, we can determine the percentage of scores lying above and below any given score value.

probabilities associated with the normal distribution

Another important property of the normal distribution is that the percentages associated with areas under the curve can be thought of as probabilities. These probabilities are percentages stated in decimal form. For example, if there is a probability that an event will take place 50% of the time, the event can be said to have a probability of .50. Likewise, an event that will probably occur 14% of the time has a probability of .14.

All of the percentages associated with areas under the normal curve can be presented in decimal form and considered as statements of probabilities associated with the areas. Figure 10 shows some of these probabilities. They are merely

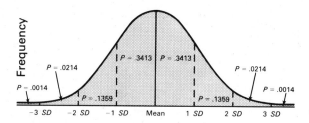

Figure 10. Probabilities associated with the normal distribution

decimal equivalents of the percentages presented in Figure 6.

The following example shows how helpful it is to consider the area under the curve in terms of probabilities. We have shown that 15.87% of the scores in a normal distribution lie above 1 *SD*. By converting this percentage into decimal form and interpreting it as a probability, we can say that the probability is .1587 of selecting an individual at random from the population who has a score of 1 *SD* or more above the mean. This figure was obtained by summing the probabilities above 1 *SD* (.1359 + .0214 + .0014). Statisticians usually write this as *P* = .1587, the *P* representing *probability*. In the same fashion, we can determine the probability of selecting at random one person who has a score that is, say, at −2 *SD* or lower. Figure 10 indicates that, for scores lying at −2 *SD* or lower, *P* = .0228. This means that the probability of selecting an individual who has a score lower than −2 *SD* is *P* = .0228, or approximately 2 in 100.

With the probability values shown in Figure 10, we can also determine the probability of obtaining a score that lies between any two given *SD* values, just as we did with the percentage values shown in Figure 8. For instance, the probability of selecting a score that lies between −1 *SD* and 1 *SD* is *P* = .6826, and between 1 *SD* and 3 *SD* is *P* = .1573.

3

introduction
to statistical
inference

Now we will learn how to use the properties of the normal distribution to make statistical inferences in research situations. Before we proceed in our statistical discussion, however, I must define some terms that are used in research.

population and sample

Up to this point we have been concerned with the normal distribution of *all* the scores of a certain description, such as all the arithmetic scores or IQ scores of a group. When we know all the scores for some group, it can be said that we know the scores for that *population*. The term *population* has a specific meaning to the statistician. It refers to certain defined measures of all people of a specific description. That is, if we have available to us the arithmetic scores of all fourth graders, we can say that we have the arithmetic scores for that population.

A population can be very small or very large. It can consist of all rocks in Yosemite Valley or all children in Mrs. Smith's first-grade class. As you may have inferred, the key word in describing a population is *all*.

When your data are taken from only part of a particular group, you have a *sample* of the population. If Mrs. Smith has 30 first graders in her class and we measure the heights of a randomly selected 10 of them, we have the heights of a sample of the children in Mrs. Smith's class. A sample is any portion of a population.

In most research situations, we cannot measure all members of the population. Instead, we usually obtain a sample and make inferences about population values from the sample values. If we are interested in the spelling scores of all fourth-grade students in the United States, for example, it would be virtually impossible to obtain scores for our entire population. Instead, we get the spelling scores of a sample of fourth graders and make inferences about the population values from these sample values.

One of the prerequisites for making accurate inferences regarding population values when using only the obtained sample values is that the sample be as *representative* of the population as possible. The most common method for obtaining a representative sample from a population is to select members randomly from the population. A random sample is one obtained by chance selection, in which each individual in the population has an equal opportunity of being selected. (There are other methods for gaining a representative sample, but these are beyond the scope of this text.)

sampling error

Now let's consider a hypothetical situation—one that would be impossible in real life but that will illustrate an important concept in statistics. Suppose we are able to obtain reading scores for every fourth-grade student in the United States. Our population therefore consists of all these fourth-grade students, since we have scores for all of them. We can then compute the mean reading score for this population. Suppose we determine that the population mean is 70. Discounting measurement errors and computational errors, we can say that the *true* mean score of this population of all fourth-grade students in the United States is 70.

Now we shall perform an experiment. Suppose we randomly select a sample of 50 students from this population and calculate their mean reading scores. We might discover that the mean of our sample is 72. If we then took another sample of 50 students, the sample mean score might be 69. It is immediately evident that, although we get sample mean scores that are close to the true mean, they are not always identical with the true mean. In fact, very seldom do they actually coincide with the value of the true mean. If we are using test scores, this variation in the values of the sample means is due partly to *errors of measurement* and partly to *sampling error*. *Errors of measurement* are those inaccuracies in test scores due to variations in the administration of the test, the physical and psychological condition of the individuals taking the test, the degree to which the test items adequately represent the totality of possible items, and many other factors. *Sampling error* is defined as the error we make in selecting random samples to represent the population. Since it is virtually impossible for a smaller group to be exactly representative of a much larger one, this sampling error is present whenever we select samples, regardless of how careful we are in randomly selecting them. It is due to chance alone.

distribution of sample means

There is an interesting phenomenon related to sampling error. If we obtained all possible samples of, say, 50 students from a given population and computed the mean of these sample means, it would be equal to the value of the population mean. And if we obtained a large series of samples and computed the mean score of each sample, we would find that these sample mean scores would be normally distributed around the population mean, whether or not the original population was normally distributed! This phenomenon is of utmost importance to the statistician, because he knows all of the properties of the normal distribution and can now apply them to the distribution of sample means.

Suppose we select from our original hypothetical population of reading scores a large number of random samples, each with 50 scores, and compute the mean score for each sample. We can then draw a normal curve representing the distribution of these mean scores. It is important to remember that this is a distribution of *mean* scores of a large number of samples and not a distribution of individual scores, such as we discussed earlier. Figure 11 presents a distribution of sample means.

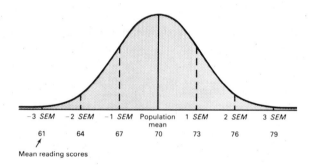

Figure 11. Distribution of sample means

You will notice that this curve has exactly the same shape as those we have previously considered. It is a normal curve, and all of the properties we have attributed to normal curves in connection with the distribution of individual scores hold true for the distribution of sample mean scores. We can see from Figure 11 that the mean of all the sample means is the population mean, and that the sample means are distributed normally around this population mean. *Remember that these samples are all of the same size.*

the standard error of the mean

We can determine the standard deviation of the sample means just as we did with individual scores. When we are dealing with mean scores, however, their standard deviation is called the *standard error of the mean* *(SEM)*, because the variability of the sample means is caused by errors of measurement and by sampling error.

In Figure 11, the *SD* units are designated as *SEM* units because this is a distribution of sample means. In our hypothetical situation, we know that the population mean is 70, and we can determine the *SEM* from our distribution of sample means around the population mean. Let's suppose that *SEM* = 3. This tells us that the standard deviation of sample means is 3 points. Knowing this, we can then determine the probability of selecting a sample with any particular mean value. If the population mean is 70, for instance, the probability of obtaining a sample whose mean is between 70 and 73 is $P = .3413$. As you can see, these determinations are made in exactly the same manner used with the distributions of individual scores.

estimate of the standard error of the mean

I said earlier that the situation in which we obtained all fourth-grade reading scores was purely hypothetical. We very seldom have the opportunity to obtain every measure in a population. It is highly impractical to do so and, furthermore, luckily for the researcher, it is not necessary. Mathematicians have developed statistical techniques for *estimating* the value of the standard error of the mean when all population values are not known. Thus it is possible to use the data in only one sample to make inferences regarding the distribution of a whole series of sample means around the unknown population mean.

Suppose we are studying the IQ scores of all high school graduates in the United States. From this population, we randomly select a sample of 50. From the data contained in our one sample, suppose we determine the following statistics:

$$
\begin{aligned}
\text{Number of scores} &= 50 \\
\text{Sample mean} &= 97 \\
\textit{SD} &= 14 \\
\text{Estimated } \textit{SEM} &= 2
\end{aligned}
$$

The *SEM* in this situation is an estimated value obtained by applying a statistical formula to the sample data. Such statistical procedures are beyond the scope of this book.

This method of estimating the standard error of the mean opens up the possibility of making inferences about population values when we know the values in only a sample of that population.

4
confidence intervals

Many research problems are concerned with trying to make an "educated guess" about how accurately a sample mean corresponds to the mean of the population from which it is selected. That is, we wish to make an inference regarding the value of a population mean using only sample data.

To attack this problem, let's first consider a population from which we can select an infinite number of samples of, say, 50 cases each. As we saw in Chapter 3, we could prepare a distribution of sample means. The mean of such a distribution would be the population mean, and the standard deviation of the sample means is called the standard error of the mean (*SEM*).

Since the means of samples are normally distributed, we can use our knowledge of the probabilities associated with the areas under the normal curve as it applies to a distribution of sample means. Figure 11, page 22, depicts this type of distribution, in which the population mean is 70 and the *SEM* is 3. Suppose we now wish to determine the interval within which 95% of the sample means lie. One of the important properties

of a normal distribution of means is that 2.5% of them lie
above 1.96 *SEM* and 2.5% of them lie below −1.96 *SEM*. The area
bounded by these two points is shown in Figure 12. Therefore,

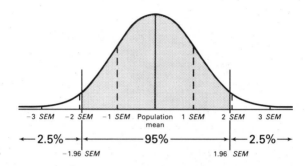

Figure 12. Interval encompassing 95% of sample means

we can say that 95% of the sample means in this distribution
lie between −1.96 *SEM* and 1.96 *SEM*. As we saw in Chapter 2,
this percentage can be considered as a probability. Therefore we
can say that, if we were to select one sample from this distribution, there is a probability of .025 that its mean lies *below*
−1.96 *SEM* and .025 that it lies *above* 1.96 *SEM*. On the other
hand, we can state that the probability is .95 that the sample
mean is *within* the interval bounded by −1.96 *SEM* and 1.96 *SEM*.
To apply this logic to our example, in which the population
mean is 70 and the *SEM* is 3, we must determine the values of the
mean at −1.96 and 1.96 *SEM*.

For −1.96 *SEM*, this value lies at 1.96 × 3 below the
mean, which is 70 − 5.88, or 64.12.

For 1.96 *SEM*, this value lies at 1.96 × 3 above the mean,
which is 70 + 5.88, or 75.88.

This interval is shown in Figure 13.

We can say that, if we randomly select one sample from
this distribution, there is a probability of .95 that its mean
will be between 64.12 and 75.88. Also, there is a probability of
05 that its mean will have a value outside (either above or
below) this interval.

The situation described above is certainly an unlikely
one, since we are practically never in a position to take an infi-

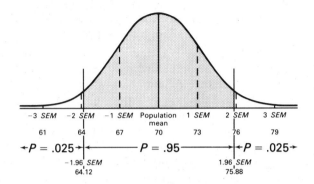

	−3 SEM	−2 SEM	−1 SEM	Population mean	1 SEM	2 SEM	3 SEM
	61	64	67	70	73	76	79

←P = .025→ |←————P = .95————→| ←P = .025→

−1.96 SEM
64.12

1.96 SEM
75.88

Figure 13. Interval encompassing 95% of sample means

nite number of samples from a population. The more typical
situation is one in which we can take only *one* sample and must
use the data obtained from it for making some inferences
regarding the population mean. For example, suppose we wish
to estimate the mean height of American males. If it were
possible to measure *all* men in the nation and compute their
mean height to be 69 inches, barring measurement errors we
could state with certainty that this represents the mean height
of American males.

On the other hand, if we are able to take only one
random sample of males and compute the mean height of this
sample, we cannot state with total confidence that the mean
of the sample is identical with the mean of the population
because, as we have seen, the means of samples vary around the
population mean; only a few samples are likely to yield means
identical with the population mean.

Of course, if we are asked to give the one value that is
our best estimate of the population mean, having the data from
only one sample, our estimated value would have to be the
sample mean. Suppose we randomly selected a sample of males
and calculated their mean height to be 68 inches. Our only
conclusion regarding the value of the population mean would
have to be that it is also 68 inches. However, we would have
little confidence that our conclusion is correct, because we
know that a sample mean is likely to be either above or below

the unknown population mean. When we use the sample mean as an estimate of the population mean, we are making a *point estimate* of a population value. It is difficult to know what degree of confidence to place in this point estimate because we have no information regarding its accuracy.

We need a procedure that will allow us to state the degree of confidence we have in an estimate. A way of doing this is to establish a *range* of values for estimating the population mean, rather than take the sample mean as a point estimate of it. That is, if we make an *interval estimate* using the sample data, we can determine the degree of confidence that our interval contains the population mean. Our procedure, then, is to establish a *confidence interval* from sample data for making inferences regarding the population mean.

the 95% confidence interval

One of the commonly used methods for making statements regarding the value of the population mean is to determine the 95% confidence interval. By using such a procedure we shall determine an interval that, if we repeated the procedure for a large number of samples, would encompass the population mean 95% of the time.

Our method for establishing the 95% confidence interval is to set the sample mean at the center of a normal distribution. The standard deviations in this distribution are labeled "standard error of the mean" because the curve depicts a distribution of a large number of sample means.

Figure 14 depicts the 95% confidence interval for the normal distribution. You will notice that this interval extends from -1.96 *SEM* to 1.96 *SEM*, because 95% of the data lie between these two points in the distribution. The probability of randomly selecting a value within this interval is $P = .95$. Furthermore, the probability of selecting a value below -1.96 *SEM* is $P = .025$, and above 1.96 *SEM* it is $P = .025$, making the total probability of obtaining a value outside the 95% confidence interval (above or below) $P = .05$.

Suppose that the measurements of our sample of American males give us a sample mean of 68 and an estimated

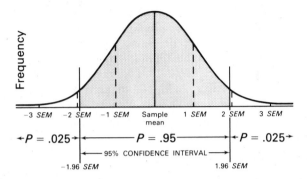

Figure 14. 95% confidence interval

SEM of 2 inches. Knowing these two values, we can draw a normal curve as shown in Figure 15, placing the mean value of 68 at the center point of the curve. With 2 points as the value of the *SEM*, we can easily determine the score value at any position on the curve. For instance, we know that value 70 will lie at 1 *SEM* because, with a mean of 68 and an *SEM* of 2, we know that 1 *SEM* above the mean is equal to 68 + 2 = 70. In this manner, we can determine the value lying at any point on the curve corresponding to any given *SEM* value.

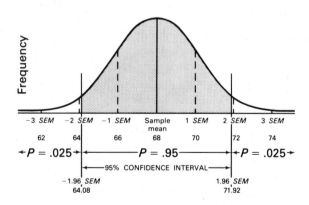

Figure 15. 95% confidence interval

Now let's determine the 95% confidence interval for
these data. We have shown that this interval is determined by
the value that lies at −1.96 *SEM* and the value that lies at 1.96
SEM. We shall first determine the lower boundary of the 95%
confidence interval—the value that lies at −1.96 *SEM*. If the mean
is 68 and the *SEM* is 2, the value that lies at −1.96 *SEM* must be
−1.96 × 2, or −3.92 points from the mean of 68. This value, of
course, is 64.08.

Now we can determine the upper boundary of the 95%
confidence interval—the value that lies at 1.96 *SEM*. In our
example, this value must be 1.96 × 2, or 3.92 points from the
mean. This value is 71.92. Therefore, the 95% confidence
interval for this distribution of data is from score value 64.08 to
score value 71.92. Interpreted in terms of probability, these
facts tell us that, for this distribution, with a mean of 68 and an
SEM of 2, the probability that a randomly selected value lies
between 64.08 and 71.92 is $P = .95$. On the other hand, the
probability that it lies outside of this interval (either above
71.92 or below 64.08) is $P = .05$. Simply stated, these facts tell
us that, if we selected values at random from this distribution,
the chances are 95 out of 100 that we would select those that
lie within the interval from 64.08 to 71.92, and 5 out of 100
that we would select those that lie outside this interval.

However, for purposes of making a statement on the
value of the population mean, it is not proper to state that there
is a probability of .95 that the population mean lies within
our established interval, because the population mean, although
unknown, is only one fixed value. Therefore, it either does
or does not lie within the interval; there is no probability
involved.

On the other hand, we can use the confidence interval
to reason that, if we could obtain a large number of samples and
determine intervals from each of them, 95% of the intervals
would encompass the population mean. Since we usually have
the data in only one sample, we would then conclude that
the probability is .95 that our interval is one of those that encompasses the population mean.

In summary, it is incorrect to say that the probability is
.95 that the population mean lies within our interval. Rather,
we should state that this process of establishing intervals, if

applied to many samples, will yield intervals that will contain the population mean 95 times out of 100.

the 99% confidence interval

If we wished to determine an interval that would give us more confidence than $P = .95$ in our statement on the population mean, we could determine the 99% confidence interval. By so doing we could state that, in the long run, 99% of the intervals derived by this method would encompass the population mean.

The 99% confidence interval is determined in a manner similar to that used to determine the 95% confidence interval. From the properties of the normal distribution, we know that 0.5% of the sample means will lie below -2.58 *SEM* and that 0.5% of them will lie above 2.58 *SEM*. In our example, in which the sample mean is 68 and the *SEM* is 2, these two values are determined as follows.

For -2.58 *SEM*, this value lies at 2.58 $\times$ 2 below the mean, which is 68 − 5.16, or 62.84.

For 2.58 *SEM*, this value lies at 2.58 $\times$ 2 above the mean, which is 68 + 5.16, or 73.16.

This interval is shown in Figure 16.

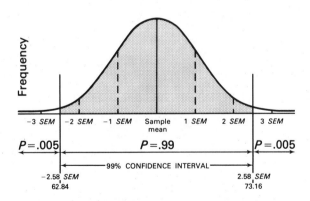

Figure 16. 99% confidence interval

It is important that the proper interpretation of confidence intervals be understood. The probability expressed in stating confidence intervals involves intervals containing the population mean. It is not a statement on the probability that the population mean will fall within a given interval, because there is only one population mean (although unknown), with a specific value, whereas there are as many possible intervals as there are possible samples. Each sample yields its own interval, and these intervals differ from sample to sample. In computing the confidence interval for a given sample, we are attempting to determine the probability that it is one of those intervals that encompasses the population mean rather than one that does not. Thus, the 99% confidence interval we determined from our sample data is interpreted as having a probability of .99 that it is an interval that encompasses the population mean.

Of course, we could determine any other confidence interval using the data in our sample. For example, we could compute the 50% confidence interval or the 80% confidence interval. I have illustrated the 95% and 99% confidence intervals because they are the ones most commonly used by statisticians.

5

comparisons among multiple samples

Up to this point we have been considering the data in only one sample. In practical research situations, we usually wish to compare two or more samples. For example, we may wish to determine whether there is a difference in achievement between students who are taught by the lecture method and those taught by the discussion method, or whether boys can jump farther than girls. Sometimes a researcher may want to make comparisons among more than two groups; but let's consider the simple situation in which there are only two groups to be compared.

Suppose we randomly select two samples of fourth-grade children, with 50 children in each sample, and teach them music appreciation by two different methods. For convenience we designate these as Method A and Method B. At the end of the school year we administer a music-appreciation test to both groups and obtain scores for both samples. We compute the mean and standard deviation for both samples and obtain the following data:

	Method A	Method B
Number of students	50	50
Mean	75	78
SD	7	8

From our data we can tell that the group of students taught by Method B (call this Sample B) received a higher mean score than the group receiving Method A (call this Sample A). If we knew that these sample means were identical to the population means for the two methods, we could indeed say that Method B was superior to Method A. But from our previous discussion we know that there is always sampling error involved when we select a sample from a population. (In this case, the population is all fourth-grade children who might be taught music appreciation by Method A or Method B.) If we took two random samples from the *same* population, the means would be different because of sampling error. The question is how much of a difference in means we need to be able to assume that the means are from *different* populations. In other words, the question that we, as statisticians, ask is "What is the probability that the difference between the two sample means is due to sampling error?" Can the difference between our sample means be attributed to random error in our sampling, or do children taught by one method actually learn more than those taught by the other method, so that we are in effect dealing with two different populations?

Before we determine how to answer this question, we must consider how a researcher makes a research hypothesis. In our example, the research question to be answered is:

> *Research question.* Is there a difference in effectiveness between Method A and Method B for teaching music appreciation to fourth-grade students?

Restated as a research hypothesis, this question becomes:

> *Research hypothesis.* There will be a difference between the effectiveness of Method A and the effectiveness of Method B for teaching music appreciation to fourth-grade students.

Here is an interesting dilemma that the statistician faces when he attempts to test the validity of such a hypothesis. First, in order to apply his statistical techniques, he must state the research hypothesis in statistical terms. For the statistician, the hypothesis might read:

> *Research hypothesis.* The mean score for Group *A* came from a different population of mean scores than the mean score for Group *B*.

The problem is that the statistician is unable to determine probabilities regarding *two* populations. Because of limitations in his theory, he can only deal with *one* normal distribution, of which he knows all of the properties. To solve his dilemma, he must find a way to restate the hypothesis in a form that can be analyzed using only one normal distribution. To do this, he rewords the research hypothesis by stating it as a *null hypothesis* (the word *null* is used because the hypothesis states that there is no difference between two things):

> *Null hypothesis.* There is no difference in effectiveness between Method *A* and Method *B* for teaching music appreciation to fourth-grade students.

To the statistician, this is the same as saying that the mean score of Group *A* and the mean score of Group *B* came from the same population of mean scores, and the difference between them is due to sampling error.

Now the statistician can apply his knowledge of the normal distribution because his hypothesis is concerned with the distribution of sample means in *one* population of mean scores.

The statistician wants to make a decision whether or not to reject the null hypothesis. He decides not to reject the null hypothesis if there is a *high* probability that the difference between his two sample means could have resulted from sampling error; he decides to reject the null hypothesis if there is a *low* probability that the difference between his two sample means could have resulted from sampling error. This is tantamount to saying that the two sample means did not come from the same distribution of sample means, and that a real difference exists between the test scores of children receiving Method *A* and those receiving Method *B*. Whether the difference results from the difference in effectiveness of the two teaching methods or from other factors, such as teacher motivation or environmental differences in the classrooms, will be a matter of concern for the researcher as he interprets the statistical findings.

Notice that the decision is whether or not to *reject* the null hypothesis. Deciding to *not reject* the null hypothesis is not the same as deciding to accept it. Why make this seemingly minute semantic distinction between "not rejecting" and "accepting" the null hypothesis? We need to remember that the null hypothesis states that there is no difference between the groups. Notice that this is an exact statement about the amount of difference between the groups. If the results of our project are inconsistent with this hypothesis, we can legitimately reject it. However, if our results are consistent with the null hypothesis, they cannot be interpreted as grounds for accepting it, because a finding consistent with the null hypothesis could be consistent with other hypotheses also. For example, in examining the null hypothesis "There is no difference between girls and boys in their problem-solving ability," if we find that girls far outscore boys, we can correctly decide to reject the null hypothesis. On the other hand, if we find that girls do not score significantly higher than boys, we cannot conclude that there is "no difference" (which would mean accepting the null hypothesis). It may be, in fact, that girls are a little better at problem solving (or that boys are a little better), but our research project failed to detect this difference. Our decision not to reject the null hypothesis means only that the data obtained from the boys and girls in our samples were not sufficiently different to enable us to conclude that it didn't happen by chance.

In summary, a researcher should not "accept" the null hypothesis; he should only "not reject" it. If a statistical test does not indicate that there is a difference between two samples, we can only conclude that, in our particular study, we failed to detect a significant difference. It does not mean that in fact no difference exists. Therefore we can only "not reject" the hypothesis that there is no difference. We are not justified in concluding that there *is* no difference.

In testing the null hypothesis, we are concerned with the difference between a pair of sample means. By stating the null hypothesis, the statistician is assuming that the difference between his pair of sample means is due to sampling error. The next step he must take is to determine the distribution of differences between pairs of sample means.

To understand how he decides whether to reject the null hypothesis, we must once again create a hypothetical situation. Suppose we have unlimited time and resources at our disposal and can take many, many samples from the *same* population. Let's assume that the number of scores in each sample is 50. Earlier we learned that a distribution of these sample means would provide us with the standard error of the mean (*SEM*). Now we are concerned with learning something about the distribution of *differences* between *pairs* of sample means. To do this, suppose we are able to form every conceivable combination of two sample means in a given population. This will give us an array of pairs of sample means. For each pair, we then determine the difference between the mean scores. Now we have an array of differences between paired sample means. If we make a distribution of these differences between paired sample means, we notice that they form a normal distribution. This normal distribution of differences between sample means selected from the same population is shown in Figure 17.

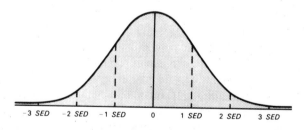

-3 *SED*	-2 *SED*	-1 *SED*	0	1 *SED*	2 *SED*	3 *SED*

Figure 17. Normal distribution of differences between sample means selected from the same population

One of the important features of this distribution of differences is that the *mean* difference score is always equal to zero. This is logical, since the samples were selected from the same population. It is evident that the differences between pairs of sample means selected from the same population would vary around zero.

the standard error of the difference between means

After we have plotted all of the differences between means, as shown in Figure 17, we can compute the standard deviation of these differences. The special term used to describe the standard deviation of differences between means is called the *standard error of the difference,* because it is a standard measure of the error involved in an estimation made on the basis of differences between two samples from a population. In Figure 17 the standard error of the difference (SED) is shown in standard deviation units, exactly as the standard error of the mean (SEM) was. The major distinction between the two measures is that the SEM is a distribution of sample means around the population mean, whereas the SED is a distribution of differences between paired sample means around a mean of zero.

From the properties of the normal curve, we know that 95% of the differences between sample means lie between $-1.96\ SED$ and $1.96\ SED$. Therefore the probability that the difference between a given pair of sample means lies between these two points is $P = .95$. Likewise, the probability that the difference between means lies *outside* these two points is $P = .05$. ($P = .025$ below $-1.96\ SED$ and $P = .025$ above $1.96\ SED$).

In our hypothetical situation, suppose that, after we have distributed all of our differences between the paired sample means, we determine that the SED of this distribution is 3 points. To determine the two values that correspond to $-1.96\ SED$ and $1.96\ SED$ in our example, we calculate as follows:

For −1.96 *SED*:   −1.96 × 3 = −5.88.
For 1.96 *SED*:   1.96 × 3 = 5.88.

Figure 18 presents the normal distribution of differences between sample means, with these two points shown as cutoff points for $P = .95$. You will notice that for these data the mean of the distribution is zero, with the differences between the means being distributed on either side of this midpoint. (The negative differences occur as a result of our method of pairing sample means, because in some pairs of sample means we are subtracting a larger mean score from a smaller one. As you would expect, this occurs in 50% of the comparisons, so that half of the distribution of differences is composed of negative differences and half is composed of positive differences.)

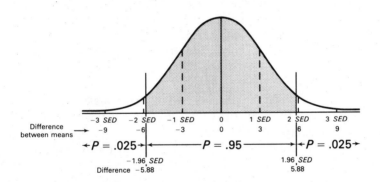

Figure 18. The distribution of differences between sample means

We can state that the probability, due to sampling errors only, of obtaining two samples whose mean scores differ by 5.88 points or less (regardless of whether the difference is negative or positive) is $P = .95$. The probability of obtaining from this population two samples whose mean scores differ by more than 5.88 points is $P = .05$.

testing the null hypothesis

The illustration of the distribution of differences between pairs of sample means (Figure 18) was presented in order to illustrate how these differences occur when we deal with an infinite number of pairs of samples and how probability statements can be made about their occurrence. In actuality, we never have the luxury of many pairs of samples; our research studies are usually concerned with only two samples. Our statistical problem in such situations is to use the values of the samples themselves to estimate the amount of variability in the distribution of differences between sample means. From this estimate we can determine the probability that the two sample means are from the same population.

Fortunately for the statistician, there is a method for making an estimate of the standard error of the difference. This estimated value is obtained by applying a statistical formula to the data in the two samples. The presentation of this statistical procedure for estimating SED is beyond the scope of this book. It is based on the size and variability of the data in the two samples.

To illustrate the process by which we make a decision whether or not to reject the null hypothesis using the data in only two samples, suppose we make the null hypothesis that there is no difference between the heights of tenth-grade boys and tenth-grade girls. This is the same as hypothesizing that the difference between the means of these two populations is zero. To test this hypothesis, suppose we measure a sample of 26 boys and a sample of 37 girls and obtain the following data:

Boy Sample	Girl Sample
$N = 26$	$N = 37$
Mean = 66 inches	Mean = 67 inches
SD = 1.81 inches	SD = 1.76 inches

These data indicate that the difference between the sample means is 1 inch. The researcher now faces two possibilities: (1) there is a difference in the mean heights of the populations of tenth-grade boys and tenth-grade girls, and this difference is reflected in the difference between the sample means, or (2) there is *no* difference in the mean heights of the populations, and the difference of 1 inch in the sample means is due only to sampling error. The statistical question facing him is: how probable is it that this difference of 1 inch is due to sampling error? As mentioned above, the statistician can use the data in the two samples to make an estimate of the standard error of the difference (*SED*) between sample means. In our example, this estimate of *SED* is .46 inch. Assuming that the null hypothesis is true, we can draw the distribution curve shown in Figure 19.

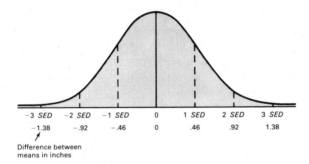

Figure 19. Distribution curve of differences between sample means

Our discussion of testing the null hypothesis mentioned that, if the probability was very small, a difference between sample means was probably due to sampling error, and we would reject the null hypothesis. Statisticians conventionally use the cutoff probability of $P = .05$ for rejecting the null hypothesis. Using sample data, we never know for *certain* whether the null hypothesis is false. All we can do is state the risk we are willing to run of rejecting it when it is really true. In using $P = .05$ as our probability level, we are saying that, if we decide to reject the null hypothesis, we recognize that there is a probability of $P = .05$ that we are making the wrong decision. Since $P = .05$ is such a small probability of incorrectly rejecting the null hypothesis, we are usually willing to run that risk. By using the data in our two samples, we can determine the probability that they came from the same population. If this probability is less than $P = .05$, we will reject the null hypothesis that they came from the same population. On the other hand, if the probability of obtaining a difference between the sample means is greater than .05, we will not reject the null hypothesis, and we will conclude that the difference between the means could be due to sampling error.

The points on the normal curve that cut off .05 of the area are located at −1.96 *SED* and 1.96 *SED*, as shown in Figure 20. Notice that the probability of .05 is split between the two tails of the distribution, *P* = .025 being in each tail.

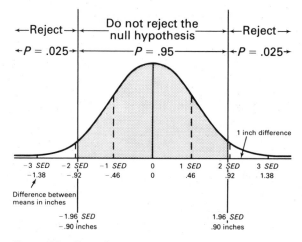

Figure 20. Areas for rejecting or not rejecting the null hypothesis at *P* = .05.

In our example, these cutoff points are at 1.96 X −.46, or −.90 inches, and at 1.96 X .46, or .90 inches. These two points indicate that there is *P* = .05 of obtaining a difference between sample means as large or larger than .90 (either plus or minus).

Having indicated the cutoff points for rejecting the null hypothesis, we now need to look back at our sample data and determine where on this curve our observed difference of 1 inch lies. Our purpose is to find out whether this difference of 1 inch falls within the rejection region of the curve. Our difference of 1 inch is indicated by the arrow in Figure 20. Since a difference of .90 inch is the cutoff point, a 1-inch difference between sample means clearly falls in the rejection region and we therefore would reject the null hypothesis. Rejection means that we are willing to state that the probability is too small (it is less than .05) and that our obtained difference between sample means was due to sampling error. We are willing to reject the null hypothesis and say that it represents a real difference between the heights of boys and girls.

If we had obtained a difference between means of, say, .5 inch, we would not have been in a position to reject the null hypothesis.

So far I have illustrated the use of the conventional cutoff probability of $P = .05$ for making decisions regarding the null hypothesis. Another commonly used cutoff probability for testing the null hypothesis is $P = .01$. This is a more stringent test because it requires us to determine the cutoff point beyond which only 1% of the differences between sample means lie. Therefore, if we are able to reject the null hypothesis using $P = .01$, the probability is *very* small (.01 or less) that our obtained difference between two sample means is due to sampling error. There is a much smaller probability of being wrong in rejecting the null hypothesis than when $P = .05$ is used as the cutoff probability.

From the properties of the normal curve, we know that there is a probability of $P = .99$ that a difference between two sample means lies between −2.58 *SED* and 2.58 *SED*. In our example, in which the estimated *SED* = .46, the two cutoff points are determined as follows:

> For −2.58 *SED*:   −2.58 × .46 = −1.19 inches.
> For 2.58 *SED*:   2.58 × .46 = 1.19 inches.

Figure 21 illustrates the areas for rejecting the null hypothesis at $P = .01$.

If you compare Figure 21 with Figure 20, you can see that the area for rejecting the null hypothesis is smaller for $P = .01$ than for $P = .05$. In using $P = .05$, we needed a difference between our sample means of only .90 inch or more in order to reject the null hypothesis. Using $P = .01$, we need 1.19 inches or more for rejection.

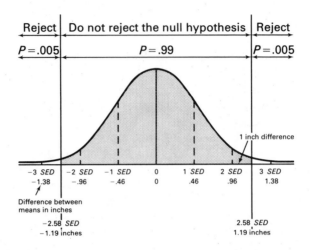

Figure 21. Areas for rejecting or not rejecting the null hypothesis at $P = .01$.

In our example (page 40), the difference between the mean heights of boys and girls was 1 inch. If we use $P = .01$ as our cutoff probability, this obtained difference lies within the region for not rejecting the null hypothesis.(see arrow in Figure 21), and we would have to conclude that the difference of 1 inch in our sample means could be due to sampling error and may not reflect a true difference in the heights of boys and girls in the population.

On the other hand, if we had obtained a difference between sample means of, say, $1\frac{1}{2}$ inches, this difference would have been in the rejection region and we would have rejected the null hypothesis, knowing that there was a probability of only .01 that we had made an error in rejecting it.

There are various ways that decisions to reject or not reject the null hypothesis are reported. A researcher may say, "The difference between the sample means was significant at $P = .01$" or "The sample means differed at the .05 level of significance" or "The null hypothesis was rejected at the .05 level." All of these statements mean the same thing: the difference between the sample means is large enough to lie outside a specified area in a normal distribution whose mean is zero.

On the other hand, if the difference between the sample means falls within the specified area of the normal curve, a researcher may report, "There was no significant difference between the sample means" or "The difference between the means of the samples was nonsignificant" or "The difference between the sample means did not exceed the .05 level of significance; therefore the null hypothesis was not rejected."

A distinction should be made here between the terms *significance* and *confidence.* The term *confidence* is generally applied to the estimation of an interval for making decisions regarding the population mean, whereas *significance* is used in reporting the level at which the null hypothesis is rejected.

6

type I and type II errors; one- and two-tailed hypotheses

If we decide, on the basis of the data in our samples, to reject the null hypothesis, we can never reject it with the *certainty* that we are correct. There is always some probability, however small, that we have made the wrong decision: that is, it may be that the null hypothesis is in reality correct and we have erroneously rejected it.

On the other hand, if the data in our samples lead us to not reject the null hypothesis, we again do not know with certainty that our decision about the null hypothesis is the correct decision. Here again, there is a probability that, in fact, the null hypothesis is false and we have mistakenly not rejected it.

The possibilities for rejecting or not rejecting the null hypothesis are diagrammed as follows.

Correct decisions regarding the null hypothesis

		Null hypothesis true	Null hypothesis false
Decisions based on sample data	Do not reject	Not rejecting the null hypothesis when it is, in fact, true. CORRECT DECISION	Not rejecting the null hypothesis when it is, in fact, false. TYPE II ERROR
	Reject	Rejecting the null hypothesis when it is, in fact, true. TYPE I ERROR	Rejecting the null hypothesis when it is, in fact, false. CORRECT DECISION

This diagram indicates the two types of errors that are possible when we make a decision regarding the null hypothesis. These are called Type I and Type II errors.

type I error

Rejecting the null hypothesis on the basis of sample data when, in fact, the samples came from the same population is a Type I error. The probability level we select for rejecting the null hypothesis indicates how much risk of being wrong we are willing to take. If we select the .05 level, we are taking a risk of being in error 5% of the time. In other words, even if the difference is found to be statistically significant, there is a probability of .05 that the obtained difference between our sample means was due to sampling error. In our decision to reject the null hypothesis we are saying, in effect, that the chances that the difference we obtained was due solely to sampling error are so small that we are willing to conclude that it was due *not* to sampling error but to a real difference between the populations from which the samples were drawn.

When a researcher decides on the probability level that he will use in rejecting the null hypothesis, he is giving the probability with which he is willing to be wrong in his decision. In the case of rejecting the null hypothesis, he is giving the probability level at which he is willing to risk a Type I error. If he selects .05 as his level for significance, he is saying he is willing to take the risk at $P = .05$ of making a Type I error. If he does not wish to take a risk that large, he may select $P = .01$

as his level for significance. At this level there is less probability of making a Type I error.

type II error

Not rejecting the null hypothesis on the basis of sample data when, in fact, the samples came from different populations constitutes a Type II error. If the data in an experiment lead us to not reject the null hypothesis when it is, in fact, false, we have made a Type II error. In such a case we have erroneously concluded that there is no difference between the populations from which our samples were selected. For example, consider the hypothesis that there is no difference between the mean heights of 12-year-old boys and 12-year-old girls. Suppose that, on the basis of measurements made of a random sample of boys and girls, we decide not to reject the null hypothesis. If, in fact, there was a difference in the mean heights of 12-year-old boys and 12-year-old girls in the population, we have made a Type II error.

In determining the probability of making a Type II error, it is necessary to postulate an alternative hypothesis and calculate the probability that the data in our experiment would lead us to not reject the null hypothesis if, in fact, the alternative hypothesis were true. In our example, in which we did not reject the null hypothesis, suppose we make an alternative hypothesis that the difference in mean height is 2 inches. There are statistical techniques that permit us to determine the probability that we have mistakenly failed to reject the null hypothesis if, in fact, the alternative hypothesis is true. This probability represents the risk we are willing to take in making a Type II error when the alternative hypothesis is true. Of course, we could have made any number of alternative hypotheses and determined, for each of them, the probability of making a Type II error.

In designing research projects, the researcher sets his level of significance at $P = .05$ or $P = .01$ *before* collecting data. Thus, he predetermines the probability of making a Type I error if the analysis of the data leads him to reject the null hypothesis. He cannot predetermine the probability of making Type II errors, but he can indirectly control them by carefully constructing his research design and by using large samples. It is, of course, desirable to keep the probability of making either Type I or Type II errors small. However, Type I and Type II errors are not independent; the more stringent you set your significance level (that is, the probability of making a Type I error), the higher the probability that you will erroneously fail to accept the null hypothesis (a Type II error).

Statistical tests differ in their ability to detect true differences using sample data when a difference in fact exists in the population from which the samples were selected. The degree to which a test is sensitive to true differences is called the *power* of the statistical test. The power of a test is reflected in the probability that a decision based upon it is a Type II error. That is, given equal probabilities of making a Type I error, a statistical procedure in which there is a small probability of making a Type II error is more powerful than one in which there is a large probability of making a Type II error.

one- and two-tailed hypotheses

At this point we should examine two characteristics of the data with which we have been dealing. One is that the research hypothesis we have been testing has stated that there is a difference between two groups or treatments; it has not stated the *direction* of the difference. That is, we have not yet made a hypothesis that specifies which group or treatment should exceed the other. (In our previous example we did not hypothesize that either boys or girls would be taller; we hypothesized only that their heights would differ.)

A hypothesis that does not indicate the direction of the expected difference but merely states that there is a difference is called a *two-tailed hypothesis.* It is so designated because it is concerned with both tails of the distribution of differences between sample means. (Recall that the distributions in Figures 20 and 21 show cutoff areas for both tails of the distribution.)

A hypothesis that states which treatment is better than the other is called a *one-tailed hypothesis* because it is con-

cerned with only one tail of the distribution of differences between sample means. An example of a one-tailed hypothesis
is "Method *B* will be more effective for teaching arithmetic to
fourth-grade students than Method *A*." Notice the difference
between this hypothesis and the two-tailed one made earlier, in
which the researcher did not hypothesize which method would
be superior. This distinction in hypotheses is important to the
statistician because in testing a one-tailed hypothesis he can use
a different cutoff point for significance than he can with a two-
tailed hypothesis.

Suppose that, in our research regarding the effectiveness
of Method *A* and Method *B*, the researcher had made the one-
tailed hypothesis stated above. In this case, the statistician would
designate all positive differences as indications of the superiority of Method *B* over Method *A* and all negative differences as
indications of the superiority of Method *A* over Method *B*. In
other words, he would be examining only one tail of the distribution of differences between means.

Figure 22 presents the same distribution of differences
between sample means as Figure 20. For our one-tailed
hypothesis, however, we can designate our cutoff point for rejecting the null hypothesis differently. Notice that the entire

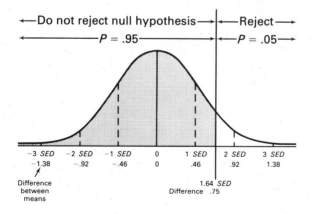

**Figure 22. Areas for rejecting and not rejecting the
null hypothesis at *P* = .05, one-tailed test.**

probability area for $P = .05$ is indicated in the right-hand tail of the distribution rather than being split between both tails, as in the earlier example.

This figure indicates that the cutoff point for $P = .05$ for a one-tailed test is .75 points. Therefore, if the mean score for Group B exceeds the mean score for Group A by .75 points, we can reject the null hypothesis that there is no difference between the methods. If the difference is less than .75 points, or if the difference is negative by *any* amount, we cannot reject the null hypothesis, and we must conclude that the difference we obtained may be due to sampling error.

The cutoff point for significance at $P = .05$ for a two-tailed test was 1.96 *SED*. We have just shown that for a one-tailed test the cutoff point for $P = .05$ is 1.64 *SED*. Therefore, a one-tailed hypothesis does not require as large a difference between mean scores as does a two-tailed hypothesis.

There are certain stipulations to when a researcher can make a one-tailed hypothesis and when he can make a two-tailed one, but these distinctions lie in the realm of research design and are beyond the scope of this book.

We have examined the basic procedure for analyzing the difference between the means of two samples. Now I shall summarize the procedure. First, the researcher develops a research hypothesis: that there is a difference between two groups or treatments. He may or may not hypothesize the direction of the difference. He also states the level of significance he will use. Then the statistician rewords this research hypothesis so that he can use it as a basis for a statistical analysis of the data. For this purpose, he states the hypothesis in null form; that is, he hypothesizes that there is no difference between the means of the two samples. This is the same as saying that whatever difference the researcher obtains between his two samples is due solely to sampling error and does not reflect a true difference between populations or treatments. Next, data are obtained from the two samples. The mean and standard deviation of each sample are computed. From these data, by use of a statistical formula, the statistician can estimate the standard error of the mean (*SEM*) for each sample. Then, again by a statistical formula, he can estimate the standard error of the difference between sample means (*SED*).

Having an estimate of the standard error of the difference, the statistician can determine whether the obtained difference between the means is significant at the stated level of significance. On the basis of this significance test, the null hypothesis is either rejected or not rejected. If it is rejected, the conclusion is that there is a real difference between the groups or treatments. The probability of making a Type I error is stated by the researcher's preselected level of significance. If the null hypothesis is not rejected, the conclusion is that the difference obtained between the means of the two samples may be due solely to sampling error and may not reflect a real difference in the two groups or treatments. The probability of making a Type II error can be determined by postulating alternative hypotheses.

Of course, the procedure described above is legitimate only if the samples were randomly selected and if the researcher can assume that the underlying sampling distribution is normal. If the statistician cannot make this assumption, he cannot properly use this statistical test. There are other statistical techniques for examining data, however, that do not require the assumption that the underlying distribution is normal.

7

t tests
and
analysis
of variance

Now let's look at another complication to our method
of analyzing differences between samples. Our examples up
to this point have used samples in which the total number of
scores was large (30 or more). The properties of the normal
distribution hold true for determining whether to reject the null
hypothesis at a specified level of significance if the samples
are large but not if the samples are small. The same decision process is used with small samples, but we have to switch to a distribution that reflects small sample size. The sampling distribution used to make statistical decisions when samples are small
tends to be flatter than the normal distribution. Figure 23
shows both the normal distribution and the flatter distributions
that occur when the samples are small. (Shapes of the curves
are exaggerated for purposes of illustration.) Notice that the distribution based upon samples containing only 15 scores is
considerably flatter than the normal distribution.

The significance of this flatter curve for the statistician
is that, for data in which the samples are small, he cannot use

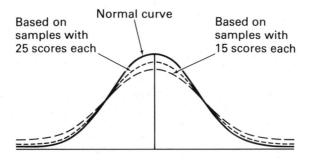

Figure 23. Comparison of distributions for rejecting the null hypothesis based on differing sample sizes

the properties of the normal curve in deciding whether to reject the null hypothesis. Instead, he must use values that reflect this flattening of the normal curve. They are called *t* values, and they have been computed to provide *P* = .05 and *P* = .01 values for samples of any given size. Statisticians have prepared statistical tables of these *t* values for all sizes of sample comparisons, so that if you know the size of each sample you are comparing you can easily determine the magnitude of the *t* value you need for significance at the .05 or .01 level.

The following table shows how a researcher might report a statistical comparison involving two samples.

	Treatment *A*	Treatment *B*
Number in sample *N*	10	17
Mean of the sample *M*	55	60
Standard deviation *SD*	4.5	4.2
t = 2.91 *P* < .01		

The last line in this table shows that the statistician computed a *t* value of 2.91. The next expression, *P* < .01, is read as "The probability is less than .01." This means that the *t* value of 2.91 indicates that the difference between the sample means is large enough to lie outside of the area for not rejecting the null hypothesis. (In this example, we needed a *t* value of 2.79 or greater to be in the "rejection" area for *P* = .01.) Therefore, the null hypothesis that there is no difference between the means is rejected, and the findings are said to be significant at less than the .01 level. In other words, the probability is less than .01 that the researcher has made a Type I error in rejecting the null hypothesis. If the *t* value had indicated that the difference between the samples was outside of the $P = .95$ area but not outside of the $P = .99$ area, this would have been indicated by the expression $P < .05$. If a *t* value indicates differences lying within the $P = .95$ area, its value is usually not reported. Instead, the researcher states that the *t* value is not significant. In this case, he does not reject the null hypothesis and attributes the difference between his sample means to sampling error.

This type of statistical test is called a *t test* and is appropriate for comparing means of small samples when it is assumed the samples have been randomly selected and the scores come from normally distributed populations. Other statistical tests are available if the assumption of normality cannot be made.

It is important to recall a point in the discussion regarding the test of the null hypothesis: to decide not to reject the null hypothesis does not mean you have *proved* it to be true but merely indicates that the observed data in your experiment are not inconsistent with the null hypothesis. If your decision is to reject the null hypothesis, this does not indicate that the null hypothesis is in fact false; it merely means that the data you obtained from your samples indicate that it is improbable that the null hypothesis is true. Deciding not to reject the null hypothesis means only that your data do not present any convincing evidence that the two samples came from different populations.

comparison of two variances

I have illustrated how the *t* test is used for comparing the mean scores of two samples. Another type of statistical analysis is available to us if we wish to determine whether the scores in one sample are more *variable* than the scores in another sample. For this comparison we use a technique called an *F test,* in which we determine whether the amount of variability in one set of data is significantly larger than the amount of variability in another. Up to this point we have been using the standard deviation as our measure of variability. However, in conducting an *F test,* we use the measure of variability called the *variance.* You will remember that we computed the standard deviation by finding the square root of the variance; the variance, therefore, is the square of the standard deviation. For example, if we know that the standard deviation of a set of scores is 4, then the variance of this set of scores is 4 X 4, or 16. The reason we must use this technique when comparing the variabilities of samples is that variances can be summed whereas standard deviations cannot.

Technically, these variances of the scores within each sample are used as *variance estimates* of the populations from which the samples are drawn. To make an *F* test between two variances, we simply divide the larger variance estimate by the smaller variance estimate. This process gives us what is called an *F ratio* between the two variance estimates. At this point, the question we want to answer is whether the variance estimate obtained from one sample is significantly larger than the variance estimate obtained from the other sample. First, we must make a null hypothesis about the comparison of two variances, just as we did about the comparison of two means. In this case the null hypothesis is that there is no difference between the variability of scores in one sample and the variability of scores in the other sample. We then use a table of *F* values, which have been computed at the .05 and .01 levels of significance, to determine whether or not to reject the null hypothesis. Such a table gives us the values of the *F* ratios we need in order to tell whether two variances differ at these levels of significance. As with most statistical tests, large samples do not require as large an *F* ratio for significance as small samples, because there is less sampling error involved in large samples.

As an example, suppose we make the research hypothesis that there is a difference in the variability of anxiety scores between freshmen and seniors. We select two random samples and obtain the following data.

	Freshmen	Seniors
	$N = 41$	$N = 50$
	variance = 33.5	variance = 74.1

The null hypothesis to be tested in this example is "There is no difference in the variability of anxiety scores between freshmen and seniors." To test this hypothesis, we compute an *F* ratio by dividing 74.1 (the larger variance) by 33.5 (the smaller variance). This gives us $F = 2.21$. The *F* table tells us that, based on the sizes of the samples, our *F* ratio must be at least 1.82 to be significant at $P = .01$. Since our obtained *F* ratio of 2.21 exceeds the *F* ratio needed for significance, we reject the null hypothesis. Our conclusion is that the anxiety scores of seniors are more variable than the anxiety scores of freshmen.

Notice that in this test we were not concerned with the *level* of anxiety scores for the two groups. (We didn't even report the mean anxiety scores.) We were just interested in the variability of the scores.

analysis of variance

Let's examine another use of the *F* test, called *analysis of variance*. Research studies often include more than two samples. Suppose we wish to compare the effects of four different teaching methods on students' arithmetic achievement. If we randomly assign students to the four methods, we could measure their arithmetic achievement at the end of the school year and obtain a mean score for each of the four samples. In this study the null hypothesis is "There is no difference in the *effectiveness* of the four teaching methods on students' arithmetic achievement."

We expect some variability among the mean scores of the four samples to be due to sampling error. Therefore we wish to ask, "Is the variability among the mean scores of the samples large enough to permit us to reject the null hypothesis?" In other words, we need to determine whether the variability among the means represents a *true* difference or whether it is likely to be due to sampling error.

The *F* test can be used to analyze the variability among the mean scores of three or more samples if we can assume that the samples were randomly selected and are from normally distributed populations. (In fact, it can also be used to compare two samples, but in that case it would be identical with the *t* test.) The *F* test used for comparison of several mean scores is called the *analysis-of-variance* technique and involves the comparison of two variance estimates.

We want to compare an estimate of the population variance obtained from scores within each sample with an estimate obtained from the mean scores of the various samples. To make our analysis-of-variance test, we are going to compare these two variance estimates to see if they differ significantly. This is the crux of the analysis-of-variance technique, because if we can show that the variance estimate based upon the mean scores of samples (the between-samples variance estimate) is significantly larger than the variance estimate estimated from the scores in the samples, we can conclude that the samples did not come from the same population. This conclusion will lead us to reject the null hypothesis.

One of the variance estimates is obtained by computing a variance estimate for each of the four samples separately, and then combining these four estimates to obtain one estimate. This process is called the *within-groups variance estimate* because it is obtained by computing the variance estimate within each sample and combining them.

The other variance estimate is computed by obtaining the mean score for each of the four samples and computing a variance estimate using these four mean scores and the size of the samples for the computation. This variance estimate is called the *between-groups variance estimate* because it is the variance estimated from the means of the various samples.

Thus we have two estimates of the population variance. We wish to determine whether the between-groups variance estimate is significantly larger than the within-groups variance estimate. To do this, we compare the variance estimate derived from the means of the groups with the variance estimate derived from the scores within each of the groups. If the between-groups variance estimate is sufficiently larger than the within-groups variance estimate, we may reject the null hypothesis and say that the samples did not come from the same population. In other words, the variability in the mean scores of the samples is too large to have occurred due to sampling error. In our example, this conclusion would mean that there was a difference in the effectiveness of the four teaching methods.

If we cannot reject the null hypothesis, we conclude that the samples may have come from the same population and the variation in the four sample means may be due to sampling error. In this case, we would decide that there is probably no difference in the effectiveness of the four methods of teaching arithmetic.

To apply the analysis-of-variance technique, an *F* ratio is computed between the two variance estimates, using the between-groups variance estimate as the numerator and the within-groups variance estimate as the denominator. By using tabled values for *F*, we can determine, for any particular sample sizes, the *F* ratio needed to reject the null hypothesis at the .05 or .01 level of significance.

As an example, suppose we wish to determine whether levels of illumination affect work production in an electronics firm. We randomly select four samples of 40 employees each and assign them to work under different levels of illumination. We then measure the work production of each group and obtain the following data:

Level I	Mean = 40
Level II	Mean = 38
Level III	Mean = 27
Level IV	Mean = 26

It is evident that the mean work production of the four samples differs. Again, we need to know whether the variability among the sample means (that is, the difference among the means) occurred as a result of sampling error or can be attributed to the amount of illumination.

In this example, the null hypothesis to be tested is "No difference in the work production of employees results from differences in levels of illumination." To determine whether or not we are justified in rejecting the null hypothesis, we examine this set of data using the analysis-of-variance technique. We interpret the results of this statistical technique in the same way we interpreted the *t* test. The analysis-of-variance technique yields an F ratio, which can be evaluated by using tables of F values prepared by mathematicians. Suppose that the F ratio in our example is significant at the .01 level. We would then reject the null hypothesis and conclude that the level of illumination is related to work production.

Inspection of the sample mean scores reveals that work production between Levels I and II differs only by 2 points. Also, the difference between Levels III and IV is only 1 point. The major difference appears to be between Levels II and III. Our analysis-of-variance technique told us only that there was an overall difference among the four means; it did not indicate which groups produced the significant difference. There are other statistical techniques that examine the amount of difference between each of these levels.

If the analysis-of-variance technique had yielded an F ratio that was nonsignificant, we would not have rejected the null hypothesis and would have concluded that the difference in work production could have been due to sampling error rather than to the effect of differing levels of illumination.

The analysis-of-variance technique may be used for analyzing the differences among any number of samples; it is also applicable for analyzing differences among groups within samples, such as male-female or age-level groupings. In our example, we could have divided our samples into male and female employees and also further divided them into age categories. Of course, in this case we would have had to have many more employees in our samples. Using the analysis-of-variance technique, we could have examined differences in work production between the sexes, according to the age of employees, and among the differing levels of illumination. This would be called a three-way analysis of variance because we would have been analyzing the work production by sex, age, and amount of illumination. It is possible for the analysis-of-variance technique to detect differences in work production due to any of these three factors or any combination of them.

8

correlation

Up to this point we have been considering sets of data in which each individual has only one score. Another form of statistical analysis is available to us if we have scores for two variables for each individual in the group and wish to determine whether there is a relationship between these variables. For instance, we have measured the arithmetic achievement and spelling achievement of a group of fifth graders and wish to know the degree of relationship that exists between these two sets of scores. In this case, we are looking not for *differences* between two groups of individuals but at the degree of relationship between two measures on the same individuals.

The statistical technique appropriate to this inquiry is *correlation*. Technically, we would say that we have scores on the variables of arithmetic achievement and spelling achievement and wish to determine the degree of correlation between these two variables. Suppose we have scores for eight students:

	Arithmetic Achievement Scores	Spelling Achievement Scores
Student A	6	6
Student B	4	4
Student C	3	3
Student D	2	2
Student E	8	8
Student F	5	5
Student G	1	1
Student H	7	7

To depict graphically the relationship (correlation) between the variables of arithmetic achievement and spelling achievement, we shall draw what is called a *scatter diagram*. To draw this diagram we choose one of the variables—say, arithmetic achievement—to be represented on the vertical axis, and the other variable, spelling achievement, to be represented on the horizontal axis, as shown in Figure 24.

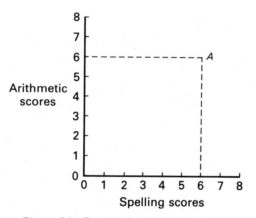

Figure 24. Format for a scatter diagram

Notice that in this diagram the range of arithmetic achievement scores is shown along the vertical axis, with the lowest score placed at the bottom, and the range of spelling achievement scores is shown along the horizontal axis, with the lowest score placed at the left. This is the conventional method for arranging the scores in a scatter diagram.

Our data indicate that Student *A* received an arithmetic score of 6 and a spelling score of 6. To plot these two scores for Student *A*, we must locate both scores in the matrix and make one dot that will represent both scores. To do this, we locate value 6 on the arithmetic axis of the diagram and extend a line from this position across the diagram. Next we locate value 6 on the spelling axis and extend a line vertically from this position. At the point where these two lines intersect we place a dot. This one dot now represents both scores for Student *A*, as shown in Figure 24.

If we follow the same procedure for each student in our group, we will have a scatter diagram of the scores for all students on both variables, as shown in Figure 25.

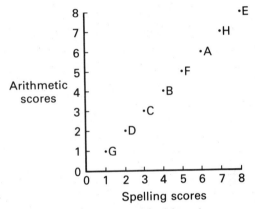

Figure 25. Scatter diagram of arithmetic scores and spelling scores

If we draw a straight line through the data in our scatter diagram, we notice that every dot falls exactly on the line. See Figure 26.

For these data it is evident that for every increase in score value on one variable there is a corresponding increase on the other variable. Since this is true for every pair of scores in our data, we conclude that the relationship between arithmetic scores and spelling scores is *perfect*. Statistically, we would call this particular relationship a *perfect positive correlation*. It is called "perfect" because the amount of increase in scores

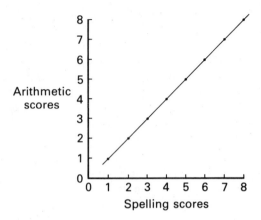

Figure 26. A perfect positive correlation

on one variable is exactly proportional to the amount of increase in scores on the other variable, with no exceptions. It is called "positive" because an *increase* in scores on one variable is associated with an *increase* in scores on the other variable.

Now let's look at a scatter diagram that depicts a perfect *negative* correlation. Figure 27 is a scatter diagram representing the relationship between the speeds of runners and the amount of weight they are carrying. It is evident from this diagram that the speeds of runners are *inversely* related to the amount of weight they are carrying. Statistically, we would say there is a *perfect negative correlation* between these two variables. Again, it is "perfect" because if we drew a straight line through the

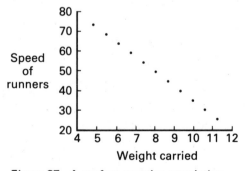

Figure 27. A perfect negative correlation

dots in the scatter diagram we would find that all dots fall
exactly on the line, and that the amount of increase on one variable is proportional to the amount of decrease on the other
variable. It is "negative" because the two variables are inversely
related.

One of the most common types of correlational statistical techniques is the *Pearson product-moment correlation*. The
degree of correlation, as measured by this particular technique,
is indicated by the size of a coefficient of correlation. The
symbol for this correlation coefficient is r.

The coefficient of correlation for the perfect positive
correlation shown in Figure 26 is $r = 1.00$. The coefficient for
the perfect negative correlation shown in Figure 27 is $r = -1.00$.
These two values are the maximum values for r. It should be
noted that the sign of the correlation coefficient indicates
whether the correlation is positive or negative. Also, the size of
the perfect correlation is the same (1.00) regardless of whether
it is positive or negative. This is an important point to keep in
mind because people sometimes mistakenly think that a coefficient of $r = -1.00$ represents no correlation. The coefficient
that indicates no degree of correlation is $r = .00$. This is the case
when scores on one variable are not related in any way to scores
on the other variable. Figure 28 presents a scatter diagram of
two such unrelated variables—IQ and rifle marksmanship scores.

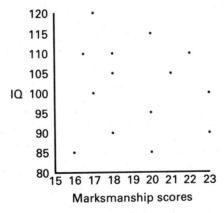

Figure 28. A scatter diagram of unrelated variables

As can be seen from this figure, the IQs of the individuals in this group have no relationship to their accuracy in shooting a rifle. The correlation coefficient indicating this lack of relationship is $r = .00$.

We mentioned earlier that correlation coefficients may be of any value from 1.00 to −1.00. As you may imagine, a perfect correlation between two variables occurs very infrequently. In almost every instance in which a relationship exists between two variables, it is less than perfect. In such cases, the coefficient is less than 1.00. For example, $r = .85$ indicates that there is a fairly strong positive correlation between two variables, $r = .54$ indicates that a less strong positive correlation exists, and $r = .03$ indicates that practically no correlation is present; $r = -.75$ indicates a fairly strong negative correlation, and $r = -.12$ indicates a very weak negative correlation. Thus we see that all positive coefficients indicate direct relationships and all negative coefficients indicate inverse relationships, and that the size of the coefficient indicates the strength of the relationship.

Let's examine closely an imperfect relationship between two variables. Figure 29 presents a scatter diagram of IQ and reading scores for a group of students.

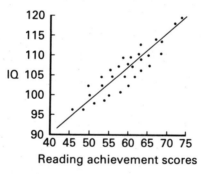

Figure 29. Scatter diagram of IQ and reading scores

By inspection we can detect that the dots tend to lie in a positive direction (that is, from the lower left to the upper right corner), although they certainly do not lie in a straight line. This tendency indicates that the correlation is positive but, since the dots do not lie in a straight line, less than perfect. Using a statistical formula, we can compute the correlation coefficient from the reading scores and IQs of the children and find that, in this case, $r = .75$. Let's look at how the size of a correlation coefficient is related to the scatter of the dots in our diagram. First we draw a straight line through the dots that best represent the trend shown in the diagram. This line is positioned in our scatter diagram so that the average distance of the dots from it is the smallest possible. In Figure 29 we have positioned a line through the dots. If we measured the perpendicular distance of each dot from this line and computed the average of these distances, it would be smaller than we could obtain if we placed the line in any other position in the diagram. A line placed in this fashion is called a *best-fit* line.

The total of the distances that the dots lie from this best-fit line is inversely related to the size of the correlation coefficient. For example, if there is a large amount of scatter among the dots, the distances of the dots from the best-fit line are large and the size of the coefficient is small. On the other hand, if there is a small amount of deviation from the best-fit line, the coefficient is large. If there is no deviation from the best-fit line (as in Figure 26 or 27), the coefficient would be either 1.00 or −1.00.

Like mean scores of samples, correlation coefficients computed from sample data are affected by sampling error. For example, if we infer the degree of correlation present in a population when we have only the correlation coefficient based on a sample, we must have a measure of the probability that our r is significantly larger than zero. That is, we wish to determine the probability that the r obtained from sample data did not occur as a result of sampling error but reflects a true relationship existing in the population. For this purpose, it is possible to compute the standard error involved in correlations. The magnitude of the standard error is in inverse relationship to the size of the sample, as might be expected. It is reasonable to assume, as in all inferences from samples to populations, that there is less error involved when you select large samples than when you select small ones. It is possible to determine the magnitude of r needed for significance at the .05 level and at the .01 level, just as we did with the *t* test. Tables have been prepared that indicate, for each sample size, the size of *r* required for significance at these two probability levels.

A word of caution is needed here on the interpretation of the meaning of a correlation. I have stated that a correlation coefficient indicates the amount of relationship between two variables. This should not be taken to mean that there is necessarily any *causal* relationship between them. Correlation does *not* imply that, because two variables are related, one is causing the other. A simple example will illustrate that correlation cannot be interpreted this way. Suppose we find a correlation between children's neatness of appearance and their punctuality in arriving at school. By no stretch of the imagination could we say that being neat *causes* the children to be on time or that being punctual *causes* them to be neat. In reality, the relationship may be caused by some third variable— such as the kind of parental attention the children receive.

Another caution should also be observed in interpreting correlation coefficients. Because they are coefficients, they cannot be interpreted as percentages of agreement. A coefficient of *r* = .30 does not indicate that there is a 30% agreement between the two sets of scores. Also, it is not proper to say that a correlation of *r* = .40 is twice as strong as a correlation of *r* = .20 just because the coefficient is twice as large. Coefficients of correlation are useful only in judging *relative* strengths of association and for indicating significant relationships between two variables.

9

chi square comparisons

So far we have been discussing the use of statistical techniques with data in the form of scores. In some studies, however, we are interested in determining whether there is a difference in the number or frequency of people responding in certain ways. For instance, we may ask a random sample of 50 people if they prefer Brand A or Brand B coffee. In this type of study we are concerned with the frequency of people's preferring either one or the other brand of coffee. We are not dealing with *scores* but with *frequencies*.

A statistical technique appropriate for data in the form of frequencies is called the *chi square* test. Using this technique, we can determine the probability that the frequencies we observe in our study differ from some theoretical hypothesized frequencies. In our coffee survey, we may have a frequency of 20 people preferring Brand A coffee and 30 people preferring Brand B. For this study, the null hypothesis is that there is no difference between the number of people preferring Brand A and the number preferring Brand B. If the null hypothesis is correct, out of 50 people we would expect 25 to prefer Brand *A* and 25 to prefer Brand *B*. In the chi square test, these are called the *expected* frequencies; that is, they are the frequencies we would expect to occur by chance.

We observed that 20 people preferred Brand *A* and 30 preferred Brand *B*. In the chi square test, these are called the *observed* frequencies. We observed that more people in our sample preferred Brand *B* than Brand *A,* but, as in all situations on which we obtain data from samples, there is a possibility that the difference between the preferences is due to sampling error and not to a true difference in the population. Our question, then, is "Are the observed frequencies sufficiently different from the expected frequencies to justify rejection of the null hypothesis?"

The chi square test, like the other tests we have considered, provides us with a statistic based on the differences between observed and expected frequencies. The test tells us whether the difference between observed and expected frequencies is significant at, say, the $P = .01$ or $P = .05$ level. On the basis of the chi square test, we determine whether the observed frequencies in our sample differ significantly from the expected frequencies based on the null hypothesis. If they do, we reject the null hypothesis and conclude that there is a preference among the population for Brand *B* over Brand *A*. If they do not, we conclude that the difference in frequencies obtained from our sample may be due to sampling error.

Chi square tests are not limited to frequencies in only two categories. If we randomly select a sample of 200 males and 200 females and ask each of them to select which of five clothing styles they prefer, we will have ten frequency counts. Suppose we obtain the following data:

		Males	Females
Style *A*		75	42
Style *B*		62	34
Style *C*		28	33
Style *D*		22	76
Style *E*		13	15
	Totals	200	200

From inspection of these data it appears that more males than females prefer Styles A and B, and that more females than males prefer Style D. Again we can use the chi square test to determine the probability that the difference between male and female preferences is due to sampling error and not to differences in preferences within the population from which the samples were selected.

Statistical tables provide the value of chi square needed to reach significance at $P = .05$ and $P = .01$. In our example, the value of chi square is 47.76. From the table of chi square values, the value we need for significance in this particular example at $P = .05$ is 9.49. Because our obtained chi square is larger than the value required, we reject the null hypothesis and conclude that the differences we observed indicate the existence of true differences in the population. If we had obtained a chi square value that was less than the tabled value we would not have rejected the null hypothesis and would have concluded that the differences we observed between males and females could have been due to sampling error.

Chi square tests can be employed with frequencies that are divided into any number of categories. The only requirement for the appropriate use of the chi square test is that the frequencies be independent of each other. That is, a frequency count in one category must in no way influence the frequency count in other categories. The value of chi square needed for significance depends on the number of categories into which the frequencies are divided.

10

conclusion

You will have noticed that the same underlying rationale is used in all the statistical tests discussed in this book. In practically all research situations, the researcher must make a decision about a population when he has information only about a sample or samples. He usually wishes to reject the hypothesis that there is no difference between his samples and to know that there is a very low probability that he has made an error in rejecting it. Statistical techniques make it possible for him to make a decision regarding the null hypothesis and tell him the probability of error in his decision.

In summary, the following are the steps involved in testing hypotheses, as discussed in this book.

1. Before he begins his research, the researcher selects the probability level at which he is willing to reject the null hypothesis. This is the probability level that he will require for significance, the highest probability he is willing to tolerate of being wrong in rejecting the null hypothesis. In effect, it is the risk he is willing to take of making a Type I error.

2. He restates the research hypothesis as a null hypothesis, which permits analysis of the data in terms of the probability that they are from a common population.
3. He selects samples, using proper sampling methods, and obtains data from them.
4. He applies an appropriate statistical test, which tells him the probability that the sample data came from a common population.
5. If the probability level obtained in step 4 is lower than his predesignated probability level for significance, he rejects the null hypothesis. In effect, he concludes that the differences among his samples reflect true differences in the populations they represent. But, even when he decides to reject the null hypothesis, he realizes there is still a small probability that he has made an incorrect decision. If the probability is higher than his predesignated probability level for significance, he does not reject the null hypothesis.

Although a researcher cannot, on the basis of sample data, be certain that he has come to the correct conclusion or proved his theory, he can, by the proper application of statistical techniques, determine with a high degree of accuracy the probability that his decision is correct.

index